ALL ABOUT
PREHISTORIC CAVE MEN

ALL ABOUT PREHISTORIC CAVE MEN

by **Sam and Beryl Epstein**

foreword by **Carleton S. Coon**

illustrated by **Will Huntington**

RANDOM HOUSE
NEW YORK

allabout books

FIRST PRINTING

 Published in New York by Random House, Inc., and simultaneously in Toronto, Canada, by Random House of Canada, Limited.

LIBRARY OF CONGRESS CATALOG CARD NUMBER: 59-6459

MANUFACTURED IN THE UNITED STATES OF AMERICA

Contents

Foreword

Sometimes there is a gap between the scientists who study prehistoric man and the authors who write books on the subject for young people. Books about Neanderthal man, for instance, often show him walking with a stoop, with his head set far forward. Actually he walked upright and held his head just the way people do today.

The authors of this book, writing about early men who lived in caves, have closed the gap. Sam and Beryl Epstein traveled to France and Spain to visit important caves once occupied by prehistoric men. They also consulted at least one person who has actually excavated caves—namely, myself.

In this book, young readers will find accurate information about a long and little-known period in the story of man. During that time, our ancestors developed many of the qualities and skills that are most important to us today.

Nobody can afford to face the future without a perspective of the past. It is never too soon to learn about who we are and where we came from. This book is a step in the right direction. I wish it well.

CARLETON S. COON
Professor of Anthropology and Curator of Ethnology
The University Museum
University of Pennsylvania

ALL ABOUT
PREHISTORIC CAVE MEN

1

What We Mean by "Cave Men"

Men have lived in caves for a long time, for one reason or another. Many still do today.

In Italy, where whole towns were bombed to rubble during World War II, thousands of people were forced to move into caves because their own homes had been destroyed.

In certain French villages, many families live in caves by choice. Their cave homes are just as snug and comfortable as their neighbors' stone houses, and much easier to build. A French cave owner doesn't have to put up four walls and a roof in order to have a house. He just puts one wall across the front of a cave, with a door and windows in it—and his house is finished. Many French farmers also use walled-up caves as barns.

Those Italian and French cave dwellers, and other present-day people who live in caves in other parts of the world, could all be called true "cave men."

But usually, when people talk about cave men, they mean the prehistoric cave men of long ago—men who lived by hunting mammoths and other prehistoric beasts.

The only weapons and tools those men had were the ones they could make for themselves out of bone and wood and stone. Their clothes, if they wore any, were probably crude garments made of animal skins. When the weather was warm, they could live in the open. But when it grew wet or cold, they sought the shelter of caves.

Not all prehistoric men lived in caves, of course. Some of them inhabited areas where no caves existed, or where the climate was so warm they could remain in the open all year around. But the prehistoric cave men lived where the climate grew so cold, at least during part of the year, that they

Some prehistoric men found shelter in caves.

needed some sort of shelter. A cave was the most convenient shelter they could find.

Not even a very large book could tell you everything about those cave dwellers. Nobody is sure exactly what they all looked like, or exactly what they all thought and felt, or how they lived. Since the cave men didn't know how to write, they left no records that we can read today—the sort of records that would tell us everything we would like to know about them. Until the last century, only a few people believed that there had been any human beings living on the earth in those long-ago times called prehistoric, or "before history."

Now, every year, clues are being found that tell us more and more about the prehistoric cave men. Some of the clues are things prehistoric men made, such as stone tools, and pictures on the walls of caves. Others are animal bones, some charred by fire. Still other clues are bones of some of the prehistoric men themselves.

Even the experts don't always agree on what a certain clue means. Take, for example, the remains of a prehistoric fire found in a cave. One expert may say it means that the men who lived there knew the secret of fire-making, and could build a fire whenever they wished. Another expert may believe that the prehistoric inhabitants of that cave had never learned how to start fires themselves. Instead, he thinks, they knew only how to keep a fire going, from year to year and generation to generation, after they had found some wood set ablaze by lightning or an erupting volcano.

Experts also disagree as to when certain kinds of prehistoric men first appeared on the earth, and when and why they finally disappeared thousands of years later. That's why books about prehistoric men, including this one, don't all give the same date for some prehistoric event.

But there are many things that the experts do agree on. They agree, for example, that a certain kind of cave dweller, known as Peking man, once lived near the great Chinese city of Peiping, or Peking.

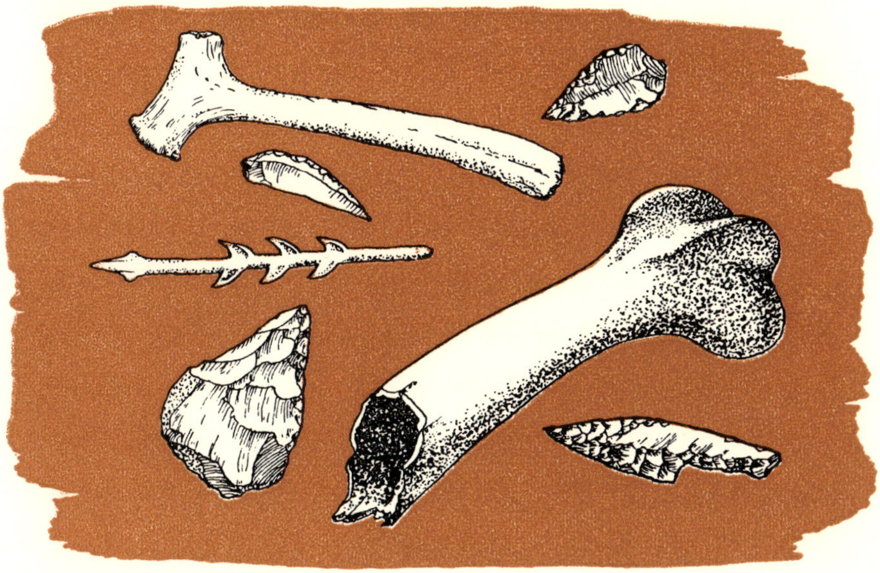

2

Peking Man

When Peking man lived in northern China, probably about 400,000 years ago, the climate of that land was a good deal warmer and damper than it is now. Some of his neighbors were the peaceful grazing animals that wandered in herds over the grassy plains of the region—prehistoric types of horses, elephants, camels, sheep, buffaloes, rhinoceroses, and small flat-antlered creatures called fallow deer. Others were huge, fierce, meat-eating wolves and bears, giant hyenas, and terrible tiger-like saber-toothed cats.

We don't know whether a Peking man's body

was covered with hair or not. We do know that his head was rather small, and that he had a very low forehead. The heavy, bony ridge that jutted out over his eyes was a good deal like the ridge over the eyes of a gorilla. He was about five feet tall, and his legs were bowed.

He probably made certain sounds that other cave men could understand—sounds that stood for such things as food and danger. But no one knows whether he had what we would call a real language.

He ate berries and other growing things, and the flesh of whatever wild game he could catch and kill. When game was very scarce, he probably sometimes ate the flesh of other Peking men.

In some ways, at least, Peking man must have seemed as much of an animal as the wild beasts around him. But he did certain things that animals never do.

He used fire, for example. Behind a curtain of flame, at night, he and his family could be safe from prowling animals in search of prey.

And Peking man made tools and weapons.

An ape sometimes picks up a handy stone, and

uses it to attack an enemy. Sometimes, too, an ape reaches for a long stick to help him knock fruit from a branch high overhead. But Peking man did more than that. He went to the trouble of *making* his tools and weapons, and he made them before the moment arrived when he actually needed them.

No ape ever looks for a stone and keeps it with him, in case he might need it. No ape starts out in the morning carrying a long stick because he might have use for it some time later in the day. Only a man thinks ahead in this way.

We don't know how or why Peking man first started to make the tools and weapons he used. But we do know that somehow he had learned to strike one stone with another until the first stone had a rough but fairly sharp edge. If the newly edged stone was big, he could use it as a sort of ax, or chopping tool. If it was smaller, he could use it as a crude but handy knife.

He did not try to make sharp-edged tools out of the pieces of soft limestone around the caves where he lived. These caves were in the limestone cliffs at the northern edge of the great Chinese plains. Instead he went to the bed of a river, at

Prehistoric men needed hard stones to make tools.

the foot of the cliffs, and searched for pieces of hard quartz. When he banged a piece of quartz with another hard stone, he could give the quartz such a sharp edge that it was a really useful ax or knife. The slivers of quartz that broke off, as he made a new tool, were probably also useful. Perhaps Peking man used the sharp points for cutting into a tough animal hide, or for splitting open a big bone so that he could get at the rich marrow inside.

Peking man did not belong to the same group or species of human beings that we belong to, the species that includes all people in the modern world. His brain, for example, probably wasn't as large as the brain of most men alive in the world today.

But his stone tools gave Peking man the strength and power to survive in a world of creatures much larger and stronger than himself. The ability to plan and make those tools set him apart from all those creatures. That ability gives him the right to be called a man.

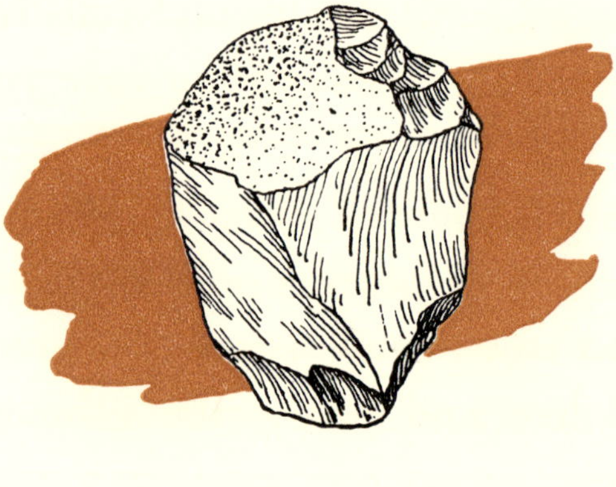

3

Finding the Clues to Cave Men

If you could walk into the cave home of a Peking man soon after he had left it, you could learn many things about him—just as you could learn a lot about some modern family if you explored their house right after they had moved out of it, leaving all sorts of odds and ends behind.

The moment you entered the cave you would know that Peking man used fire, because you would see bits of burned wood and perhaps some charred bones in a black heap on the dirt floor. Bloodstained stone tools would tell you how the cave man killed and cut up the animals he ate.

Peking man used fire.

Discarded animal bones would tell you, if you were an expert, what kinds of animals were most common in his diet.

You probably wouldn't find any hint about the color of the cave man's eyes or hair. But if you found some human bones you could begin to figure out what kind of skeleton a Peking man had —whether he was long-legged or short-legged, bigheaded or small-headed, tall or short.

Today, of course, nobody can walk into a cave man's home that has been recently deserted. In fact, today, it is usually very difficult even to find the places where cave men once lived. The chief reasons for this are the great changes that took place on the earth after man first made his appearance, probably more than half a million years ago.

More than half a million years ago the period called the Ice Age began. It ended only about 11,000 years ago. Four times during that long period great sheets of ice spread slowly down, from the North Pole and from the tops of high mountains, to cover huge areas of Europe, Asia, and North America. At places the ice was as much as several miles thick.

Each of those glacial periods, as they are called, was followed by thousands of years of warmer weather. And during those warm spells, when the ice sheets melted and shrank, the level of the ocean was raised all over the world. Then the moisture in the air increased, and there were more frequent fogs and rains.

During the Ice Age, deserts turned into green

forests, and then turned once more into sandy wastes. The faces of cliffs wore away. Rocks and soil washing down hillsides filled up whole valleys, and then streams carved new channels through the new valley floors. Rivers shifted their courses or dried up, and new rivers were formed.

During that long period, changes took place in and around many of the caves where prehistoric man had lived. The roofs of some collapsed, sealing off their entrances so that to this day no one can tell where they were. Other caves, in the path of flooding rivers, were filled up with layer after layer of silt, packed down until it turned as hard as rock. Dust, blown by the wind, drifted into still other caves, century after century, until a cave entrance perhaps fifty feet high became a shallow slit so small that no one could squeeze into it.

Now an ancient cave man's home is usually discovered only by chance, perhaps when a landslide occurs and exposes an old cave opening, or when the workers in a stone quarry break through into a long-hidden cavern.

By now most of the stone tools made by the prehistoric men have been buried deep under lay-

In the Ice Age, glaciers reached far down from the north.

ers of earth or debris, or carried by rushing streams to some spot far distant from the place where their makers lived. Even if one of them happens to turn up in a farmer's field, during spring plowing, it can't tell us very much, by itself, about the man who made it.

By now, too, the bones of the men who lived in prehistoric times have mostly been destroyed, be-

A cave man's home may be discovered by chance.

cause bones are so much more fragile than stone.

And if the early cave men made things out of woven reeds, or other very fragile materials, those things have long ago crumbled to dust, or decayed into mold, and left no trace at all.

That's why some facts about the cave men will probably never be known. No clues can ever be found to give us even the slightest hint about them.

And the things we do know about prehistoric men have all been learned only after years of work and study.

First the clues have to be found—and we have already seen why that is difficult.

Then the clues have to be recognized—and that is not always easy either. Usually only an expert can tell that a bit of bone was once part of the skeleton of a prehistoric man, or that a sharp-edged piece of stone was once a cave man's knife. Such an expert is called a paleontologist. If a farmer picks up a sharp-edged stone in his field, he is likely to toss it aside without ever thinking that it might once have been used as a knife by some prehistoric man.

Then, finally, many clues have to be fitted together in the right way, so that they tell us a true story of part of the mystery of life on earth in the Ice Age.

Sometimes years go by between the finding of a clue and the moment when it is fitted into place, as part of the solution of that still-mysterious time long ago.

That's what happened after the finding of what proved to be the first clue to Peking man.

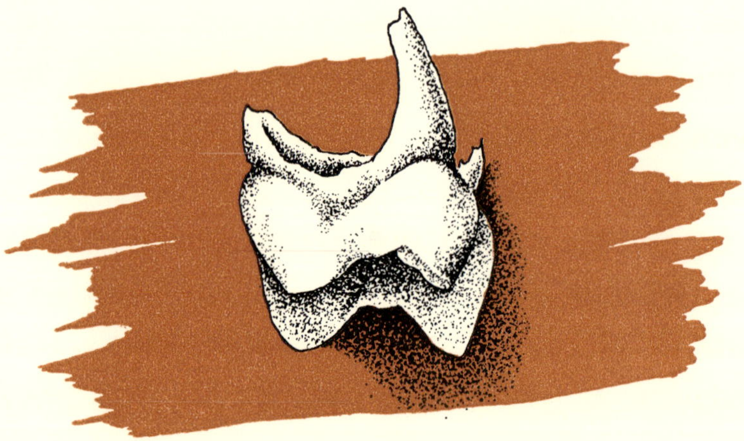

4

The Strange Story of a Tooth

The first clue to Peking man was one single tooth, and for many years no one even recognized that that's what it was.

In the year 1899 that tooth was lying, along with a lot of other ancient teeth and bones, in the little shop of a Chinese druggist. The shop owner always kept on hand all the old bones and teeth he could get. Perhaps he bought them from farmers who picked them up in their fields. But he told his customers that the bones and teeth had once belonged to ancient fire-breathing dragons. Ground up into a powder, he said, they made a powerful

The druggist had old bones and teeth.

medicine. Like other druggists all over China at that time, he had many customers who were willing to pay a high price for that kind of "dragon" medicine.

One day a German came into the shop, saw the bones and teeth, and said he would buy them all. He didn't believe that dragons had ever existed. He didn't think "dragon" medicine could cure anything. But he had noticed that the old teeth he saw in Chinese shops looked too large to have belonged to animals of the kind now living on earth. He had decided that they must be the teeth of prehistoric animals. He thought they would be valuable to paleontologists, the scientists who study ancient forms of life. That's why he stopped at every Chinese drug store he passed, and bought all the old bones and teeth he could afford. He planned to take them home as a gift to a friend who was a paleontologist.

His friend was delighted with the present. He and the other scientists of that day still knew very little about prehistoric man, but they had been finding clues to prehistoric animals for a long time, and already knew quite a lot about them. So the paleontologist went happily to work, studying his new collection of bones and teeth. In 1903 he wrote a long report on them. In it he described almost a hundred prehistoric animals which—the bones and teeth proved—had lived in China thousands of years ago. Other scientists became so excited when they read the report that many of them went to China in the hope of collecting more of the same kind of animal bones and teeth for their own collections.

But certain scientists were especially interested in one tooth among all the hundreds bought in the Chinese shops. They didn't think it was an animal's tooth—unless, they said, it had belonged to some prehistoric ape that no one yet knew anything about. They thought it might have belonged to a prehistoric man. And if it was the tooth of a man, they said, the man had lived much longer ago than any of the few other prehistoric men to whom clues had already been discovered.

They asked the German who had bought the tooth exactly where in China it had been found. They hoped to go to the same spot themselves and look for more clues of the same kind. But of course he didn't know. So nobody went to China to look for clues to a prehistoric man, because nobody had any idea where in all that vast country they should start their search.

After a time the strange tooth was almost forgotten. The man who first began to think about it again, more than twenty years after its discovery, was a Swedish geologist, J. Gunnar Andersson.

Andersson had gone to China to help Chinese geologists study their country's mineral wealth. His job was to inspect mines and quarries. He also enjoyed looking for fossils, as a hobby. While he studied the mines and quarries, he found a good many old bones. After he had looked at them himself, he marked each one to show exactly where it had been found, in case some scientist wanted to explore that same spot again. Then he sent all his finds to a Swedish university where they could be studied by professional paleontologists.

In 1920 Andersson was inspecting some limestone quarries at a place called Choukoutien, not far from the city of Peking, now called Peiping, in northern China. He noticed that there were veins of reddish sandy soil in the limestone cliffs of the neighborhood, and he immediately became excited. Those veins, he thought, might once have been holes, or caves, in the pale limestone—caves which over the centuries had become filled in by reddish wind-blown soil. And if they had once been caves, he thought, they might contain, imbedded in the red soil, all sorts of very old animal bones. He set workmen at the job of digging out the sandy reddish stuff.

His guess turned out to be correct. When the soil was dug away, it was easy to see that there had been caves in the limestone—caves littered with the bones of prehistoric animals. Andersson gathered up all those bones too and sent them off to Sweden.

But as he packed up the bones, he noticed that some of them had been broken or crushed in a curious way. Had they been smashed by the weight of the sandy soil blown in upon them? He didn't think so. Had they been chewed up by some large beast? He didn't think that was likely either. If they had been crushed by the weight of soil, or chewed up by a beast, he thought they would look different. He began to wonder if the bones might have been broken by a club or stone held in the hand of some prehistoric man.

Then, soon afterward, he noticed something else. Among the bones were pieces of sharp-edged quartz. He wondered how the quartz had found its way into limestone caves. He knew that quartz and limestone are not naturally found together. He knew the quartz must have been brought to the caves from somewhere else.

Andersson became even more excited. Animals, he knew, didn't carry rocks or stones from one place to another. But if animals hadn't carried the quartz into the caves, and if it hadn't got into the

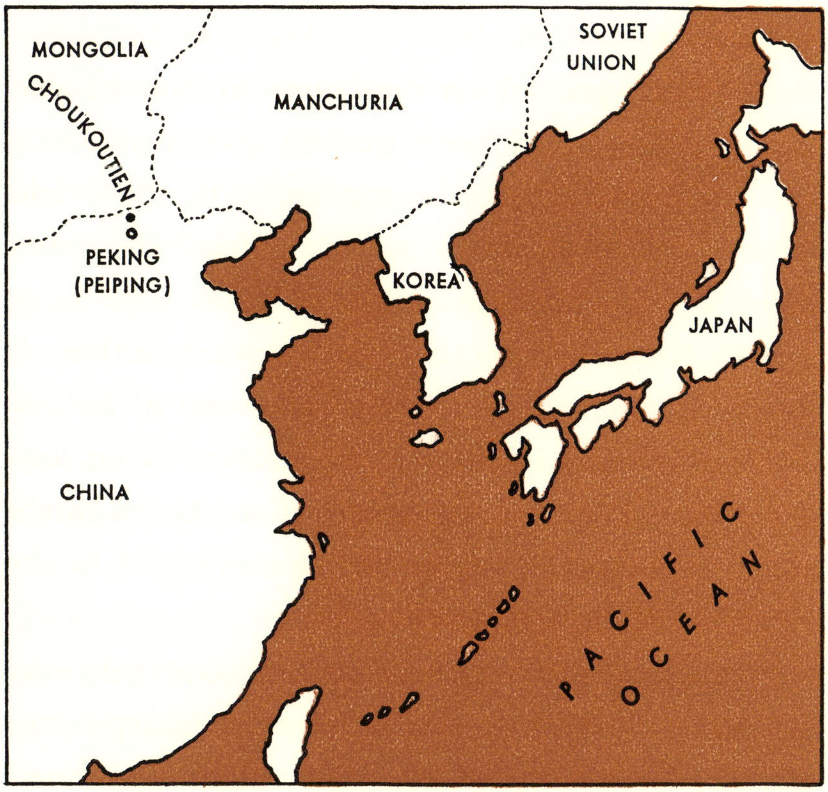

caves by some natural means, then men must have brought it there—at some long-ago time when the caves were becoming filled up with reddish soil.

Now, Andersson told himself, he had two reasons for thinking that men must have occupied the limestone caves at the time when prehistoric beasts were alive. The first clue was the oddly crushed bones. The second clue was the quartz.

That's when he remembered the tooth he had heard about, which had been taken to Germany from China and which had puzzled the experts. Was it possible, he asked himself, that that tooth had really belonged to a human being, and that it had come from one of the very limestone caves where he was working, at Choukoutien?

Andersson couldn't answer his own questions. But he thought answers might be found if the ancient Choukoutien caves were explored by an expert in the study of prehistoric man. He couldn't afford to hire such an expert himself, but fortunately he knew a man who could, a man who took a deep interest in science. So Andersson wrote to a wealthy Swedish industrialist, Ivar Kreuger, explaining his ideas and why he thought the Choukoutien caves might prove to be important to the whole world. Kreuger, impressed by Andersson's letter, sent the expert Dr. Otto Zdansky to China.

Zdansky went quickly to work, and soon he had found plenty of prehistoric animal bones. Then, to Andersson's great delight, he found two teeth which looked very much like human teeth, and which also resembled the tooth that had been bought years before in a Chinese drug store.

When Zdansky's discovery was announced in 1926, paleontologists all over the world suddenly became interested in Choukoutien. Soon experts from several countries, including China itself, arrived at the caves to join in the search. A large force of men was put to work blasting and digging out tons of the hard reddish earth. They sifted all of it carefully to make certain that no tooth or bit of bone slipped by unnoticed.

Another tooth was discovered in 1927.

Dr. Davidson Black, head of the international group working at the caves, was convinced by then that man had been living at Choukoutien in prehistoric times. In honor of the big city nearby, Dr. Black called him Peking man. Newspaper and magazine stories were soon making Peking man famous in every corner of the globe.

Other experts, Dr. Black knew, thought a few teeth, some crushed bones, and some bits of quartz were not enough clues to prove the existence of Peking man. But Dr. Black wasn't worried. He was certain that more clues would turn up, and that they would all point to the same thing.

Sure enough, in 1928 Dr. W. C. Pei, Chinese member of the group, found a piece of human skull, a human jawbone, and more human teeth. And the next year Dr. Pei settled the matter once and for all when he dug up the top of another skull—a piece of bone big enough to serve as a very definite clue to the shape of Peking man's head.

Even then, of course, the search didn't cease. Instead it went on faster and more enthusiastically than ever. Soon more bones were dug up. Along with them were found pieces of stone that Peking man had clearly chipped into the shape of useful tools. The patient scientists even found tiny seeds, amazingly preserved, that told them Peking man had varied his diet of wild animal meat by eating berries.

Bones of early men were found in caves near Peking.

In 1937 Japan invaded China, but even the war that followed didn't stop the work at Choukoutien. Nearby Peking remained in Chinese hands, and the growing collection of Peking man's bones seemed safe there. By 1941, however, the situation had grown much more threatening, and it was decided to send the bones to the United States for safety.

Early in the morning of December 5, 1941, the bones left Peking under the protection of a detachment of United States marines hurriedly departing for the United States. Two days later, before the marines could board the American ship waiting for them on the coast, the Japanese attack on Pearl Harbor brought America too into the war. The marines were captured before they could leave China. Everything they had with them disappeared into enemy hands.

The priceless bones of prehistoric Peking man, which were hidden in the marine commander's personal baggage, have never been seen again.

Did the Japanese soldiers throw the bones away when they looted the marines' baggage? Were the bones transferred to a ship bound for Japan, a

Some scientists think Peking man looked like this.

ship which was then torpedoed and sunk? No one seems to know.

Fortunately plaster models had been made of most of the bones. The models had already been sent out of China for the benefit of scientists in other lands. Those models were still safe. They are still being studied today by scientists in various countries.

Those scientists fit fragments of skull bone together, first in one way and then in another. Then they make models of a whole skull, and cover the models with clay, in the shape of the muscles and flesh they think a Peking man might have had. In this way, from the models of those teeth and bones, they are learning more every year about the human beings who lived in Choukoutien about 400,000 years ago.

The "dragon" tooth bought in a Chinese drug shop in 1899 started a long story that is still not finished to this day.

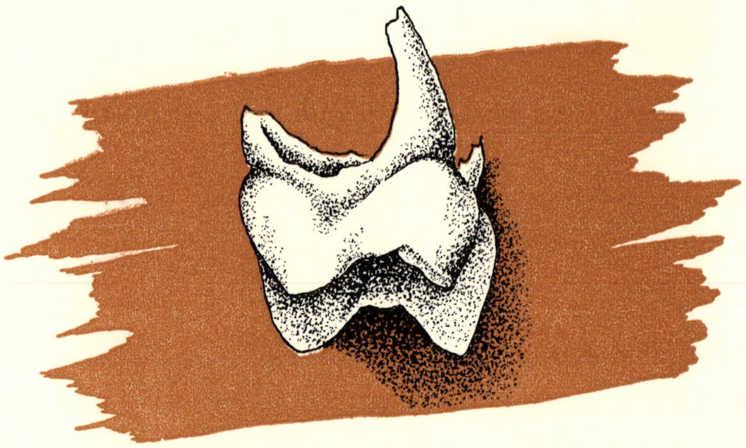

5

Neanderthal Man

Clues found in caves have also told us a good deal about another and later prehistoric cave dweller, Neanderthal man.

Beginning about 140,000 years ago, before the last of the four great ice sheets spread southward, men of this particular kind inhabited many parts of the world for at least 70,000 years. We know they were living in southern Europe, northern Africa, and the lands east of the Mediterranean. Perhaps they were living in still other places too, where no clues to them have yet been found.

These people are called Neanderthals because

Many Neanderthal families lived in caves.

the first bones of this particular kind of prehistoric men were found in a small valley in Germany called the Neanderthal.

Like Peking man, a Neanderthal man was likely to be fairly short, probably only an inch or two over five feet in height. His arms and legs were stubby. His hands and feet were short and wide. His neck was short too, but his face was long and he had a big nose. His chin sloped backward below his mouth, and his forehead sloped backward above the bony ridges that stuck out over his eyes.

But the Neanderthal man was very different from the Peking man in one way: he had a brain as large as the brain of modern man.

When people of the Neanderthal type first appeared in Europe, the climate there was fairly warm. But slowly, over thousands of years, as the fourth great sheet of glacial ice began to spread, the air grew colder and colder. It warmed up again for a time, when the ice melted a little. Then it turned cold once more when the ice spread southward a second time, chilling the air ahead of it for hundreds of miles.

The kind of animals the Neanderthals hunted

changed as the weather changed. When the climate was warm, they hunted animals that were at home among the forests of oak and beech and walnut trees that sprang up during that period—wild horses and oxen, wild boars, big red deer, and bears. When the weather turned colder, and the forests were replaced by barren tundra, those animals fled toward the south. Then the Neanderthals hunted the cold-loving animals that came into the region—polar foxes, for example, and reindeer and huge mammoths.

Probably each catch was the signal for a huge feast for the group of Neanderthal families that lived and hunted together. Carrying a burning brand from the fire which was always kept going in their small settlement, the men, women, and children all gathered around the spot of the kill. There they built a great blaze and roasted the beast. When they had eaten their fill of meat and of the soft rich marrow found inside the big bones, they carried home anything that was left.

A Neanderthal hunter had to be very brave, and very skillful too, to bring down a giant mammoth. It was easier to kill a big cave bear, especially

Neanderthal hunters banded together to kill a huge beast.

during the winter when those animals crawled into a cave for their long hibernation. Then, if a man's family needed food, he could go into a bear cave and slaughter one of the drowsy beasts with little trouble, almost as a modern farmer might go to his pig pen to slaughter a pig.

Perhaps this explains why at least some of the Neanderthals had a special feeling about bears. We know they did, because we know they collected great numbers of bear skulls in certain caves, and arranged them neatly in piles. Sometimes they put the skulls in box-like containers made of big flat stones. Apparently they even held ceremonies or rituals around the bones.

Did those Neanderthal men worship bears? Did they think bears were sacred, as Hindus today think of cows as sacred? Did they, in other words, have a religion in which bears played an important part? No clues have yet been found which give us a definite answer to such questions.

But there are certain clues which do suggest that the Neanderthals believed in a life after death. Those clues are human graves containing skeletons surrounded by tools and other objects. Apparently the Neanderthals thought a dead person would live again, and would need his tools and weapons in his new life.

We may never be able to learn whether the Neanderthals had a real religion or not. But we already know a great deal about them as makers of tools.

Probably, like most primitive men, they used pointed sticks, hardened in a fire, as digging tools or light spears for killing small animals. But for sharpening the sticks, and scraping the flesh from bones, and for all the many other things they had to do in order to keep alive, Neanderthal men needed other tools and weapons. They made them out of stone, usually out of the kind of hard stone called flint, which can be chipped to a sharp edge.

The earliest Neanderthals made only a few varieties of tools, and most of those were large. But gradually their tools became more delicate. The two made most often by the Neanderthals, probably because they found them most useful, were a sharp-pointed triangle and a kind of knife with one long sharp edge.

The sharp point of the triangle could be used to

Neanderthals made sharp-edged scrapers of stone.

slash meat or skin, or to cut tough tendons. The same triangle could also be fastened to a long wooden shaft, to make a spear. Of course, no one knows exactly what the Neanderthals' spears looked like, because the wooden shafts long ago disappeared.

The other sharp-edged stone, usually called a scraper, could be used to scrape pelts, or to cut up almost anything. Probably the scraper too was sometimes fitted into a handle.

6

Making Stone Tools

Today, in museums all over the world, there are collections of stone tools made by prehistoric men. If you see one of these collections, you might think at first that most of the tools were just ordinary stones that you could pick up in a field. But if you look more carefully you will see that each one is shaped in a special way.

And if you try to make a stone tool yourself, out of a small rock—if you try to make a scraper, let's say, shaped to fit neatly into your hand, and with a good sharp edge—you may give up pretty quickly. Your fingers will probably be hit as often as the stone, and your face may be cut by flying stone splinters. You will probably agree with the experts when they say that a cave man had to have skill and knowledge and patience in order to be a good tool maker.

Today nobody can watch a Neanderthal tool maker at work, transforming a piece of flint into a well-shaped scraper, or into a point for a spear shaft. But a French stonecutter, who became curious about prehistoric men, decided to try to copy their tools. He has helped us to understand how tool makers must have worked, in that long-ago period sometimes called the Stone Age.

The French stonecutter's materials were the same hard brittle materials the cave men preferred —quartz, obsidian, pebbles, and flint. He never worked with a steel hammer, or any other metal object, because he knew the cave men had no metals of any kind. For shaping his tools he used nothing but pieces of wood, bone, or stone. After years of study and practice, he could produce every kind of Stone Age tool ever found. He showed that the cave men probably used several different working methods, depending on the kind of tool they wanted to make.

One of those methods has been called the flake method.

To make a tool by the flake method, the tool maker starts with a piece of stone much larger than the tool he wants to make. Let's say he decides to start with a large piece of flint, the most commonly used of all the cave men's materials.

He may find a stone of the very size he wants lying on the surface of the ground in certain places. Or he may dig it out of a limestone cliff. Pieces of flint are often found, embedded in rows, in limestone.

Flint itself is usually brown or yellowish or dark gray. But a piece of flint that has been exposed to the weather for many centuries has a whitish coating on the outside. The tool maker may remove all the white coating first, because it is not as hard as the dark inside of the stone. Or he may leave the white coating on the part of the stone that will not have to be sharpened.

When he is ready to start work, he strikes his

piece of flint with a striker. This is a piece of sandstone, or another piece of flint, or a piece of hard wood such as oak or box wood. If he is not careful, he will shatter the brittle flint into many pieces with his first blow. But if he knows just how to strike it, near the edge, he will knock off a small piece called a flake.

This flake is the basis for many different kinds of tools. If many tiny flakes are then chipped from one side of it, giving it a sharp edge, it becomes a scraper. Trimmed on both sides, and given a sharp tip, it becomes a spear point.

Sharpening an edge is not always done by striking it. Sometimes it is done by pressing very hard with a stick, perhaps using a stone to tap the other end of the stick for extra pressure. This is called pressure flaking.

The other most important tool-making method used by the Neanderthals, and by many other cave men before and after their time, has been called the core method.

To use this method, the workman strikes a piece of flint time after time, in such a way that many flakes are knocked off from around its edge. He is not interested in the flakes he knocks off, although a good Stone Age workman probably saved some of them and found them useful. Instead he is interested in the original stone itself—or, rather, in the core of it that is left when he has removed the flakes. Stone cores were also the basis for many different kinds of tools. A heavy one, rounded at one end to fit into the hand, could be used as an ax or chopper.

If a workman made a certain kind of round core, by knocking off flakes in a certain way, he was ready to make a tool by a third method called the blade method. If his core was correctly made, and he gave it a single blow at the right spot, he could strike off a flat, blade-shaped piece of flint. That single blow, in other words, gave him an excellent stone knife.

Most cave men used both the core method and the flake method. Only the later tool makers, living after the time of the Neanderthals, developed and used the more difficult blade method.

Some of those later cave men, wonderfully skillful, used their stone knives and chisels to carve delicate bone needles and bone harpoons with

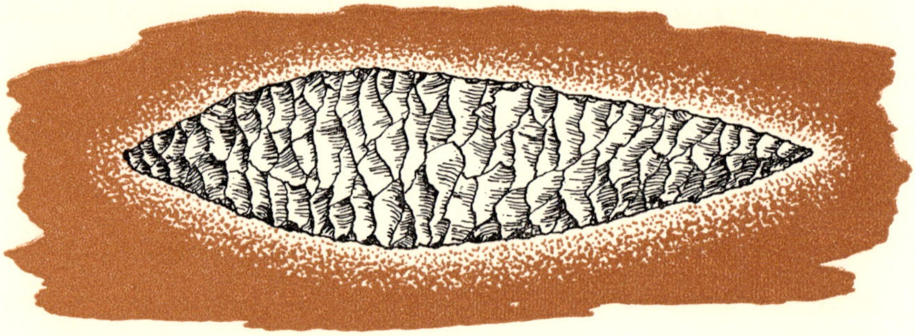

Some prehistoric craftsmen made delicate blades of stone.

curved prongs. These were tools that no earlier men had been able to make. Sometimes, too, the later cave men made a leaf-shaped stone blade so fine and delicate that, unless it was handled with great care and skill, it would have broken the first time it was used. Perhaps a blade like that was made to be used in some kind of ceremony—the way a silver sword is sometimes presented today to a general, as a mark of honor, though nobody expects him to use such a sword on a battlefield. Or perhaps that delicate blade, usually called a laurel-leaf blade, was made by a man who simply enjoyed looking at beautiful things. He may have liked to use his own hands to create an object that was beautiful to look at.

Other good clues as to how the cave men made and used their tools have come to us from people who still live very much as the cave men lived. Today most Stone Age people, as they are sometimes called, have disappeared. But in Australia, and in parts of South America, there still exist tribes that have never learned to use any kind of metal. If someone gives a Stone Age Australian a steel ax today, he may be glad to have it. But when he makes his own tools, he makes them out of wood or bone or stone.

Scientists who have gone to Australia to study the primitive tribes there learned that a Stone Age Australian could make a useful tool by the core method in a few minutes. A tool made by the flake method usually took him much longer. One man was seen striking off flake after flake, and throwing each one away because he wasn't satisfied with it. He threw away three hundred stone flakes before he finally struck off one that pleased him.

The scientists also found that a Stone Age Australian always goes to a certain special place, rather like a modern quarry, to get the stone for his tools.

Usually he makes his tools right on the spot. Probably the ancient cave men also had favorite places where they went for their tool material, and often made their tools right at their quarries.

The very early cave men used up large quantities of flint or other material, because the tools they made were often large, and because they spoiled many pieces of stone before turning out one good ax or chopper. This is why the early prehistoric men preferred to live in areas where they could find plentiful supplies of material for their stone tools, as well as plentiful game.

But the later cave men were better workmen and made smaller tools. They were also skillful at putting a new edge on a blade if it broke or became dull. This meant they could carry home enough flint or other material to supply them with tools for months. That's why the later cave men could live pretty much where they pleased, and began to wander into parts of the world where no men had ever lived before. Their tool-making skill had given them a freedom the earlier cave men didn't have.

7

Tools Are Clues

When we look at prehistoric stone tools in a museum, we can see that in a general way the tools became better as time went by. Tools made by Peking man, about 400,000 years ago, don't look as well made as Neanderthal tools chipped into shape only about 100,000 years ago.

This doesn't mean that all the people living in the world at one time had the same degree of skill, or that they all gradually improved at the same rate. One group of cave men might be making excellent stone knives, while another group living at the same time might still be making knives that were heavy and awkward to handle.

One group of cave men might also have a style or fashion of tool making different from that of another group. And the two groups might not make the same kinds of tools. One group might use the core method more often than the flake method, and make only a few stone scrapers but many stone axes. The other group might prefer the flake method, and make a great many sharp-edged scrapers. Perhaps the people of this group had found they could use the scrapers in many ways.

Each group made the tools it found most useful, and made them in its own way. When a modern expert studies a whole collection of tools made by a certain group of cave men, he is able to learn quite a lot about that particular group's habits and customs. He can learn something about the group's special way of life, which a scientist calls its *culture*.

That's why it is very fortunate that cave men were not neat. They didn't sweep out their caves. They just left animal bones and broken tools lying about in the dirt near the entrance to their cave

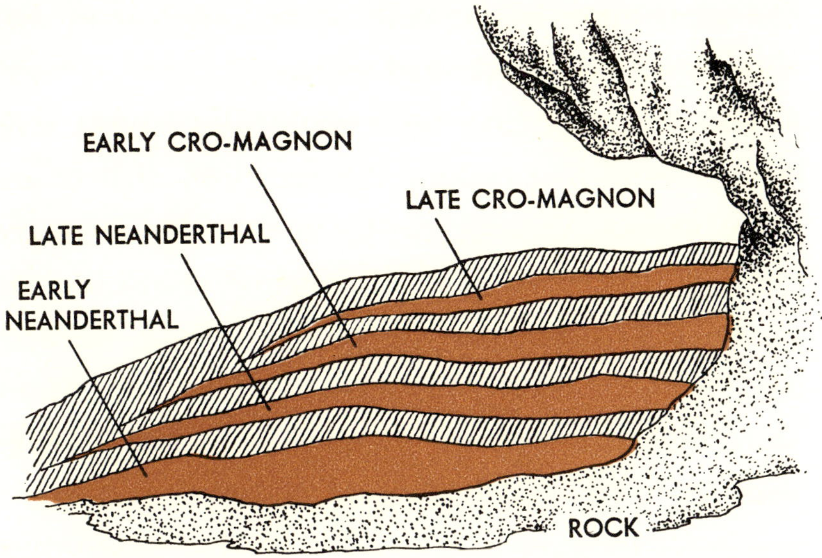

Different groups of cave men left layers of rubbish.

homes, where they spent most of their time. And if they moved away from a cave, perhaps after living in it for many, many generations, they left behind a thick layer of the litter that had been piling up there for centuries.

That litter would be covered up eventually, if the cave was empty for hundreds of years, by a layer of wind-blown dust, crumbling bits of rock from the cave roof, or the mud left by a flooding stream. Then, if other people happened to settle in the same cave, they too would begin to build up a thick hard-packed layer of litter. That layer too would be slowly covered up after those people moved off in search of new hunting grounds or a better climate.

In this way, many caves became filled up with layers of litter and rock and earth. And today's scientists, studying those layers one by one as they dig down through them, can read the history of the cave and the different groups of people who lived in it.

The entrance to one cave in northern Spain—a cave called El Castillo, or The Castle—was almost completely closed when scientists first decided to explore it. But they found that the hard-packed stuff filling the opening, which looked like ordinary earth, was really twelve different layers of cave men's litter. The layers of litter were separated by layers of soil and rock.

Carefully they dug down through the layers, one by one. Each layer of litter contained clues to one kind of cave-man culture, or way of life. The twelve layers proved that twelve different groups of people, each with its own habits and customs, had lived in that same cave at twelve different periods

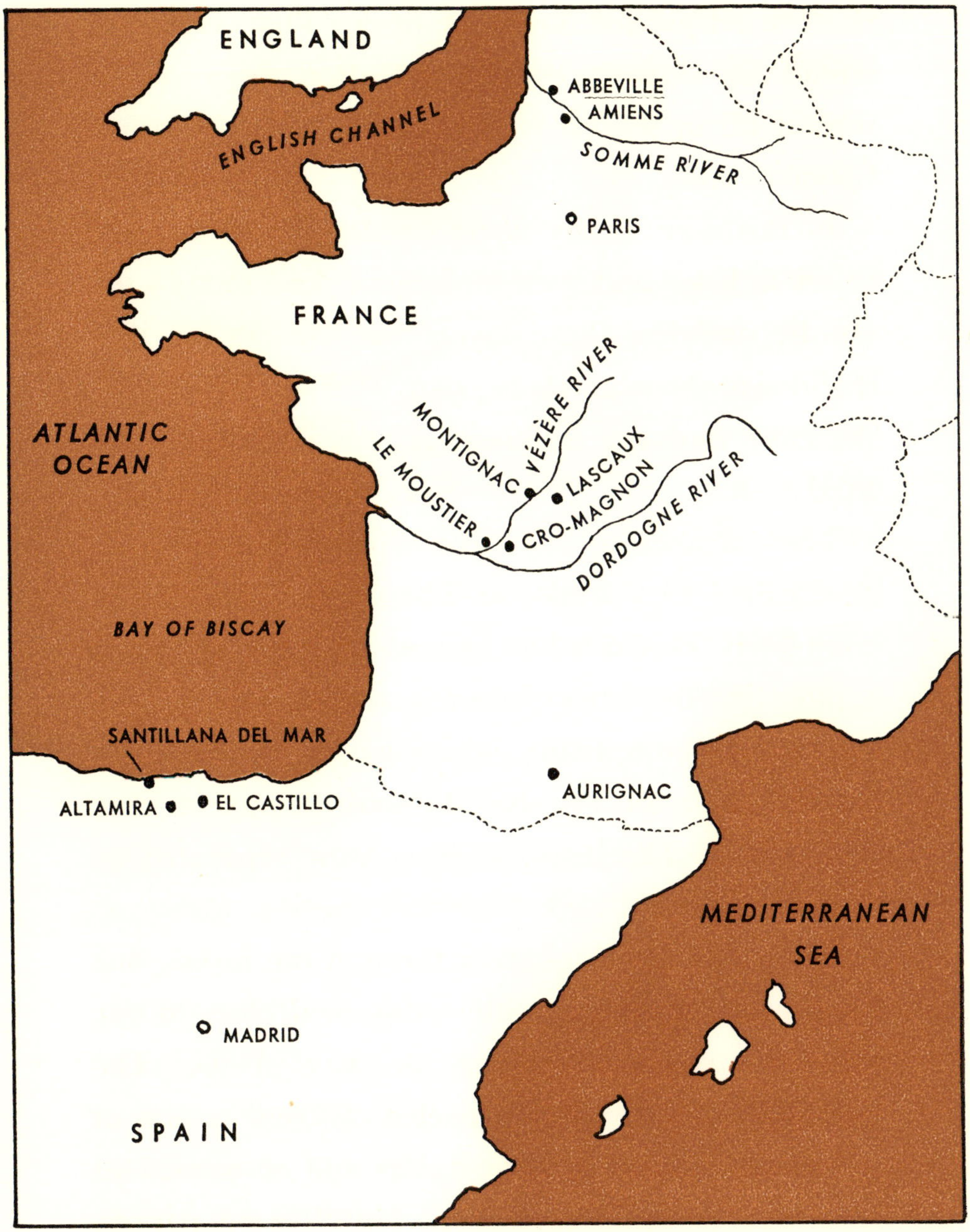
ENGLAND
ENGLISH CHANNEL
ABBEVILLE
AMIENS
SOMME RIVER
PARIS
FRANCE
ATLANTIC OCEAN
VÉZÈRE RIVER
MONTIGNAC
LE MOUSTIER
LASCAUX
CRO-MAGNON
DORDOGNE RIVER
BAY OF BISCAY
SANTILLANA DEL MAR
ALTAMIRA
EL CASTILLO
AURIGNAC
MEDITERRANEAN SEA
MADRID
SPAIN

of time. The earliest use of El Castillo as a cave home was early in the Ice Age. The cave's latest tenants probably left it as recently as ten or fifteen thousand years ago.

Scientists have given names to the prehistoric cultures they already know quite a lot about. Usually the names come from the places where clues to that particular way of life were first found. And the names—Abbevillian, for example, and Aurignacian and Mousterian—may seem hard to remember. But many French boys and girls can rattle them off very easily, because most of the names come from the names of French towns or caves.

The name of the culture called Abbevillian comes from the town of Abbeville in northern France. Aurignacian comes from a cave called Aurignac in southern France. Mousterian comes from Le Moustier, in southwest France, where a collection of tools of a certain kind were first found. Among them were finely shaped scrapers, and many stone triangles with one or two sharpened edges, all made by the flake method.

Now, whenever scientists discover a group of tools very much like the tools found at Le Mous-

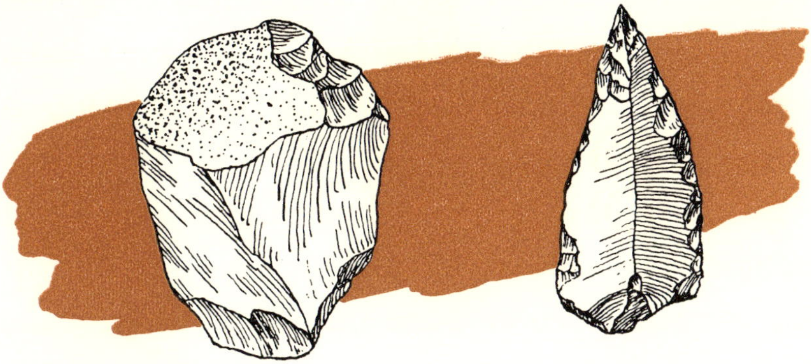

Earlier tools (left) were cruder than later tools (right).

tier, they say that the people who made them had a Mousterian culture. The tools found with the bones of Neanderthal men show that the culture of those people was Mousterian. And when scientists discovered similar Mousterian tools in many places in Europe, Africa, and western Asia, they knew that the inhabitants of a large part of the world were all living, for a time, in very much the same way.

8

A Stubborn Hero

Many different kinds of people turn up clues to the mystery of prehistoric men.

Some are people who make their discoveries by accident, perhaps while plowing a field, working in a stone quarry, or digging the foundations for a building.

Others are scientists who find new clues because they are looking for them. Museums and universities sometimes give scientists the money they need for an expedition to some distant place where it seems likely that new clues might turn up.

Searching for clues can be very disappointing.

Men can spend long months patiently scraping away at the earth without finding anything of importance.

Searching for clues can be dangerous, too. An explorer must sometimes swim through an icy underground stream, with little hope of aid if he is injured there far beneath the earth's surface. Or he may have to be lowered by a rope through a narrow twisting opening between sharp rocks. And if the rope breaks he may fall to his death.

In spite of all the difficulties, the search for new clues goes on, and new ones are found every year. And whether an important discovery is made by a farmer or an engineer, a schoolboy or an expert, the news is reported in newspaper headlines and in radio and television bulletins. All over the world people want to know more about prehistoric men. The person who makes a discovery is honored as a hero of science.

In the last century, things were different. In those days, most people simply laughed at anybody who claimed he had found bones of a prehistoric man, or tools prehistoric men had made. They refused to believe him, because they didn't believe there had been any men on earth at the time the prehistoric animals were alive.

Most people, for example, refused to believe a Frenchman named Jacques Boucher de Perthes, when he said that he had found tools made by prehistoric men. But De Perthes was very stubborn, and he kept insisting he was right. For years he refused to back down, no matter how much people laughed at him. He carried on one of the most famous arguments in the history of science.

De Perthes had not always been interested in science. As a young man with a comfortable income, he traveled a great deal. Later he became a customs inspector in the town of Abbeville, in northern France. By then his favorite hobby was the writing of plays, novels, and poems. He joined an amateur science club, but only because it was fashionable to belong to that sort of organization in his day. Among the subjects the club members discussed were the two fairly new sciences of geology and paleontology. So, without being particularly curious about them, De Perthes did know a little about both.

He knew that the earth's crust was made up of

many different layers of rock and soil, and that each layer had been formed at a different period of the earth's long history. Each layer, according to geologists, represented one period of geologic time.

De Perthes also knew that paleontologists used those layers in their study of prehistoric animals. When they found the fossil bones of two different kinds of animals in the same geologic layer, they could be quite sure that the two kinds of animals had lived at about the same time.

One day De Perthes stood watching a dredge at work in the Somme River near Abbeville. In a pile of dredged-up sand, he noticed some large bones and some odd-shaped stones. He picked them up and looked at them.

The bones, he thought, might be the bones of prehistoric animals. He decided to show them to Casmir Picard, a friend of his who was a doctor and who knew a great deal about paleontology.

The stones were interesting too, De Perthes thought, as he turned them over in his hand. Their sharp edges made them look like tools or weapons —spear points, perhaps, or crude knives. He wondered if somebody had deliberately chipped the edges of the stones to make them sharp.

Suppose, he said to himself, these bones are really prehistoric animal bones. And suppose these

De Perthes found odd-shaped stones near large bones.

stones were chipped into the shape of tools by some man. Then—if the stones and bones were lying in the same geologic layer, before the dredge unearthed them—it would mean that there were prehistoric animals and tool-making men living on earth at the same time.

That was the moment that changed De Perthes's whole life. He was sure he had made an important discovery. All the scientists of France would be amazed and delighted, he thought, if he could prove to them that there had been human beings during the time when prehistoric animals were alive. Even the greatest French scientists, he told himself, the men who had been elected to the French Academy, would say that he had made a great contribution to human knowledge.

But De Perthes knew he would have to be very sure of his facts before he announced his discovery. If he showed animal bones to a famous scientist, he would have to be sure they were genuine prehistoric bones. If he showed chipped stones, he would have to be sure they had been chipped by man's hand, not by some accident of nature. And, to prove his theory, he would have to be sure the bones had been chipped into shape at the same. time the animals were alive.

De Perthes decided to undertake a search for unmistakable prehistoric animal bones, and unmistakable man-made stone tools, lying together in the same geologic layer.

After that he spent much of his time searching the countryside and studying his finds with his friend, Dr. Picard. Farmers, quarry workers, and other laborers in the region heard that De Perthes would pay them for chipped stones or ancient bones they brought to him. So they too began to look for such things. Some of them even chipped sharp edges on stones, in the hope of earning a few extra pennies. Often these faked tools fooled the two men. Everyone around Abbeville also learned that De Perthes would pay very well if he could be shown chipped stones and ancient bones lying side by side in a layer of earth that could be proved to be very old.

Time after time, De Perthes explored the gravel pits and quarries around Abbeville. Time after time, when a new building was going up, he was

Some of De Perthes's neighbors made fake tools.

on hand to study every inch of newly exposed rock and earth. Soon the cupboards of his. house were bulging with sharpened stones and prehistoric bones. He wrote a book about some of his finds. But the final proof he sought was still lacking.

Years went by. Casmir Picard died, and De Perthes carried on alone.

In 1844, more than ten years after he began his

search, De Perthes went one day to watch workmen dig the foundation for a new hospital. That was the day he finally found what he had been seeking—found it himself.

In one geologic layer, in earth that had been undisturbed for thousands of years, he saw part of the great tooth of a prehistoric elephant, and four stones that he felt certain had been chipped into tools by the hand of man.

There could no longer be any doubt, De Perthes believed, that men and prehistoric animals had lived at the same time.

He wrote out a report of what he and Picard had done, and sent it off to Paris, to the members of the French Academy. Those men were among the most famous scientists of their day. Their theories, especially their theories about prehistoric times, were accepted by most other scientists all over the world.

Eagerly De Perthes waited for some of the great men of Paris to come to Abbeville, to see for themselves his proof that men too had existed in prehistoric times.

But none of the scientists came. According to

their theories, De Perthes could not be right. And they were certain they knew more about prehistoric times than the customs inspector of Abbeville.

In the opinion of the famous French geologists, De Perthes probably couldn't tell one geologic layer from another because he wasn't a trained geologist. Therefore, they said, he had no right to claim he had found man-made tools in a layer dating back to the time of prehistoric animals. Besides, they thought, if De Perthes really knew a great deal about stones, he would know that nature chipped them into all sorts of shapes. Only a fool, they said, would call a stone a tool just because it had a sharp edge.

The leaders of the new science of paleontology thought De Perthes couldn't possibly know much about ancient bones, either. Besides, they said, they had been finding prehistoric animal bones for years, but no well-known paleontologist had ever claimed that he had found prehistoric human bones. And that alone, they thought, could be taken as fairly definite proof that prehistoric man never existed.

And so the long argument began. De Perthes

and a few supporters stood on one side. The most respected scientists of France stood on the other.

The customs inspector of Abbeville continued to collect old bones and chipped stones. Over and over again, he wrote reports of his finds and printed them at his own expense. He sent them out in the hope that some important scientist would read them and come to see the collection he was so eager to show.

Some of his reports finally found their way to England, where scientists were beginning to explore certain new ideas that the members of the French Academy frowned upon. In 1858 and 1859, several outstanding British scientists decided to cross the English Channel to visit De Perthes.

When they reached Abbeville, the famous Englishmen carefully examined the hundreds of sharp-edged stones and countless prehistoric animal bones that filled De Perthes's house. Then they went out to the places where De Perthes had made his finds. There they dug patiently through layers of earth.

Next the British visitors went to Amiens, thirty miles away, where one of De Perthes's few supporters had also been digging for prehistoric animal bones and man-chipped stones. There too the British scientists did a thorough job of checking.

Then they returned home to write up reports of what they had seen. A few months later, when the British reports reached France, the experts of Paris were amazed by what they read.

Jacques Boucher De Perthes, the British scientists reported, had certainly made mistakes in geology. Some of the stones in the Abbeville collection had been shaped by nature and not by man. Others were fakes, chipped within the past few months or years by De Perthes's neighbors.

But, the British reports stated firmly, there were ancient hand-chipped stone tools in the collection too. Those tools proved definitely that men had been living on the earth at the time when mammoths and other prehistoric animals were alive. De Perthes, the British scientists declared, had made a major contribution to science.

Many of the more stubborn French experts were still hoping to find flaws in the British reports, or trying to ignore them completely, when they were staggered by another blow. This one was struck by a man much more modest and shy than De

Perthes. But he too was French, and he too was stubborn in his own quiet way. He was Edouard Lartet.

Lartet had been a lawyer before he gave up that career in order to explore the mysteries of the past. Like the British scientists, he was open-minded and not afraid of new ideas. He was one of the first men in France to believe that De Perthes might be right. For several years, even before the British reports, Lartet had been quietly digging

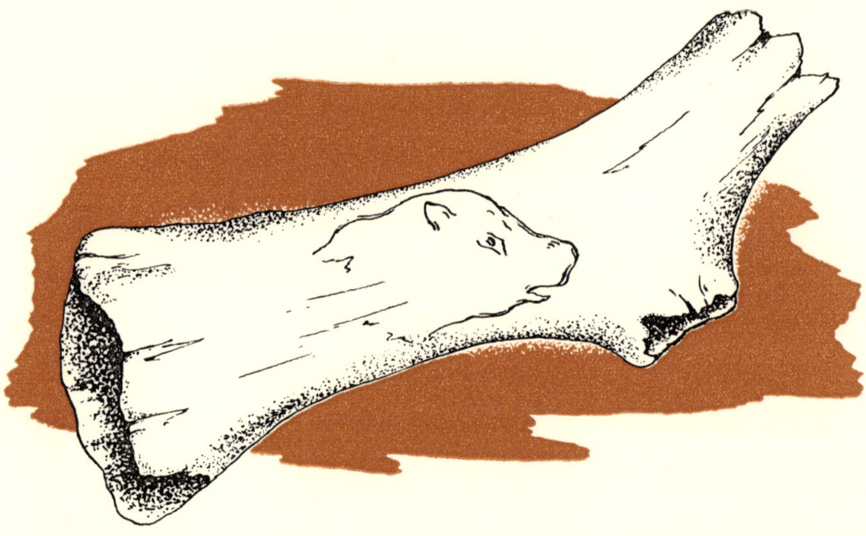

Lartet found a carving on a prehistoric antler.

about in French caves, and beneath overhanging rocks, seeking proof for the existence of prehistoric man.

In 1861 he found it—the antler of a prehistoric reindeer on which had been carved the head of a prehistoric cave bear.

The reindeer antler had been buried deep in the earth, in a layer of soil that had not been disturbed for thousands of years. There could be no doubt that the antler itself had once belonged to an Ice Age animal.

There could be no doubt that the carving had been done before the antler was buried.

There could be no doubt that the man who did the carving knew just what a cave bear looked like. And only two kinds of people knew what those bears looked like. First, people who were alive before the cave bear had become extinct, thousands of years ago. Second, the paleontologists who had recently discovered the bones of cave bears.

No paleontologist had made that carving, buried in an ancient layer of soil.

A prehistoric artist carved a picture on a mammoth's tusk.

So there could be no doubt that the little carving had been made by a prehistoric man.

Not many years later, in 1867, another remarkable find by Lartet held the place of honor at the great International Exhibition in Paris. There, in the vast exhibition hall, thousands of people crowded around it day after day.

It was only a small piece of ivory, part of a mammoth's tusk. But on it had been carved the picture of a mammoth.

"Look!" people whispered in awe, as they stared

at it. "The man who made that carving really knew what a mammoth looked like. When he was alive there were still mammoths on earth! He was a prehistoric man!"

The long argument which Jacques Boucher de Perthes had started was finally over. Prehistoric man had at last been accepted into the great family of mankind. And the new science of prehistory —the scientific study of prehistoric man and the way he lived—was about to be born.

9

Five Skeletons

What did prehistoric man look like?

That was the question some people had been asking even before the Paris Exhibition of 1867.

Now many scientists all over Europe, including those who had once argued against De Perthes's whole theory, were eager to find the answer to that question. But they couldn't find the answer without first studying some of the clues which to this day are very rare—the actual bones of prehistoric man.

Already several such clues had been discovered.

But few people had paid much attention to them, or had recognized them for what they really were.

Ten years earlier, a German schoolmaster, Johann Fuhlrött, had found part of a skull bone. It was very thick, and had a heavy ape-like ridge of bone above the eye sockets. When Fuhlrött said he believed it had belonged to some prehistoric man, almost nobody took him seriously.

Now, at the time of the Paris Exhibition, many scientists were taking a long serious look at that skull bone. But most of them still didn't believe it had been part of a prehistoric man's skeleton. Prehistoric man, they said, had certainly looked very much like modern man. They were sure he hadn't had an ape-like bulge above his eyes. And ordinary people, who knew nothing about science, were glad to agree with this point of view. Hardly anybody wanted to believe that human beings could have had ape-like ancestors. Many years passed before that bone was accepted as part of the skeleton of a Neanderthal man.

In the meantime the search for prehistoric human bones was going on in many lands. This was especially the case in the southwestern region of France, where Edouard Lartet had found that piece of mammoth tusk.

That region, called the Dordogne, has many rivers. They run through narrow valleys hemmed in by limestone cliffs. One day in 1868, a crew of road builders were blasting away part of a cliff at Cro-Magnon, a tiny village on the Vézère River. As the cloud of dust settled, after one of their blasts, they saw a black hole that looked like an opening into a cave. They all crowded around to peer inside. What they saw there appeared to be bones—not animal bones, but human skeletons.

Some of the men stood guard over the newly opened cave, while others ran with the news to Les Eyzies, the slightly larger village nearby. The officials of Les Eyzies hurried to the spot. Their neighborhood was already famous among scientists. They hoped another find had been made which would make it more famous still. They took one look into the cave, and sent off word of what they had seen to Paris.

This time none of the great scientists of Paris laughed. Even the French government took the

In a cave at Cro-Magnon were prehistoric human skeletons.

news seriously. The minister of education asked the well-known Edouard Lartet to go to Cro-Magnon immediately.

Lartet too was excited by the report, but he didn't feel well enough to travel to the south of France again. He sent in his place his young son, Louis, who had been assisting with some of his father's researches.

Louis Lartet was astonished at what he saw at Cro-Magnon. He reported to his father, to the French government, and to the scientists in Paris that the cave contained five complete human skeletons. He said he was quite sure the skeletons had belonged to prehistoric men.

The other experts who rushed to Cro-Magnon to study the bones hoped Louis Lartet was right. People who knew nothing about science hoped he was right too, because the men whose skeletons were found in that cave were the kind of ancestors they could be proud of.

Those cave dwellers, it was quite clear, hadn't looked like apes at all. Their skeletons proved that they had looked like modern men and had had full-sized brains. They had been tall, six feet in height or more. They had had long arms and long straight legs, high foreheads and firm chins. Everybody agreed they must have been very handsome.

Finally the experts gave their verdict on the skeletons of the Cro-Magnons, as such cave men are called in honor of the village where their bones were first found. That verdict is still accepted today.

Experts today agree that the Cro-Magnons belonged to the species known as *Homo sapiens.* The first word of the name is the Latin word for "man." The second word is Latin for "reasoning, intelligent, wise." *Homo sapiens* is the same species to which all people in the modern world belong.

Today some scientists believe there may have been members of the *Homo sapiens* species in Africa, and perhaps in other parts of the world, at least 500,000 years ago. Very little is known about the earliest members of this species. Perhaps they were as primitive as Peking man and other very early prehistoric people.

But the Cro-Magnons, and other groups of *Homo sapiens* very much like them, who lived at the same time, were the most advanced of all the Ice Age men we know anything about today. The Cro-Magnons inhabited Europe during the last of the four great glacial periods. Probably they reached that continent from Africa and western Asia about 70,000 years ago or more.

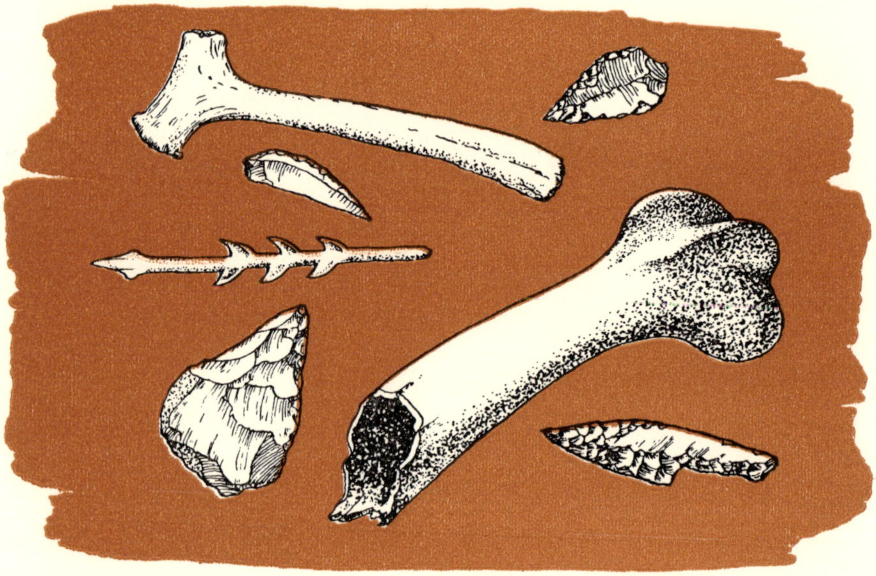

10

The Cro-Magnons

The Cro-Magnons were skillful tool makers. For example, they made harpoons, with sharp curved points, which they could use for killing reindeer.

They made spear throwers, which helped them to bring down game from a great distance. Now they could kill swift animals that could outrun a man.

They made sharp little stone chisels, which they could use to carve a piece of bone or antler into a fine needle. With that needle, Cro-Magnon women could sew animal skins into snug fur garments. In warm weather the Cro-Magnons probably wore

To kill game at a distance, the Cro-Magnons used spear throwers.

very little, painting their bodies red with iron oxide, a substance found in the soil. But their fur garments helped them to survive in bitterly cold weather.

One place where the Cro-Magnons lived, generation after generation for thousands of years, was the Dordogne region of France. Even today people who visit that region can see that it must have been a paradise for cave men.

The Dordogne region is really one big flat plateau, slashed by its many deep narrow river valleys. Moist winds blowing in from the Atlantic Ocean, not far away to the west, made the grass grow green and thick on the plateau during the Ice Age. Then countless herds of animals gathered there to graze—mammoths, wild horses, and reindeer.

The narrow river valleys held the warmth of the sun. The clear streams supplied the cave men and their wild game with plenty of water.

Cro-Magnons built fires in front of their caves.

The cliffs of the valleys are made of limestone, and are full of flint. This meant that the Cro-Magnons of the Dordogne never had to go far to get material for their tools.

Overhanging ledges of rock along these cliffs made comfortable shelters, which the Cro-Magnons used as their homes when the weather was too cold for camping on the open plateau. When they also made use of the deeper caves of the region, they lived only at the very front of them, near the entrance.

The little terrace in front of a cave or a rock shelter probably served a Cro-Magnon family as a sunny porch on a bright day. Sometimes they enlarged a terrace by edging it with stones. They built their fires there, and ate their meals alongside the warm and protecting blaze. There the women sewed, and made their necklaces out of shells and animal teeth. There the men made their tools and the other objects their families needed. Sometimes they chipped away at a big joint bone in order to fashion it into a kind of shallow bowl. Filled with animal fat, and with a bit of dry moss to serve as a wick, the hollow of the joint could be used as a lamp.

The steep cliffs and narrow valleys of the Dordogne region were very useful to the Cro-Magnon hunters. A lookout stationed on top of a cliff could see a herd of wild horses galloping over the flat plateau while the animals were still far away. He had plenty of time to call some of his neighbors to the plateau, and to station others in the valley below. Then the men on the plateau drove the animals over the cliff. The men in the valley killed the injured beasts before they could drag themselves to safety.

If the lookout saw a herd of reindeer gathering to drink at the river, in the valley at the foot of a cliff, he could summon his neighbors to the spot. Then, working together as a team, they could drive the animals toward the narrow end of the valley. And there, caught in that natural trap, the reindeer were unable to escape the spears the hunters hurled at them.

The Cro-Magnons apparently knew how to construct real traps too, and may have used them for catching very large beasts. But their favorite game

Reindeer were the favorite game of the Cro-Magnons.

was reindeer. One reason for this was that reindeer were very plentiful in Europe at that time—so plentiful that the period is sometimes called the Age of the Reindeer. Another reason was that the reindeer was useful to the Cro-Magnons in many ways.

Reindeer meat was good to eat, and very nourishing. Reindeer skins could be sewn into smooth warm garments, or cut into narrow strips for binding a spear point to its shaft. Reindeer antlers could be carved into all sorts of useful and beautiful objects.

Some of the best clues to the ways the Cro-Magnons lived are the bits of carved reindeer antler found in the Dordogne and other regions. A harpoon, or spear thrower, decorated with a carved design, shows that those cave men enjoyed making their tools and weapons beautiful.

Cro-Magnons sometimes decorated their spear throwers.

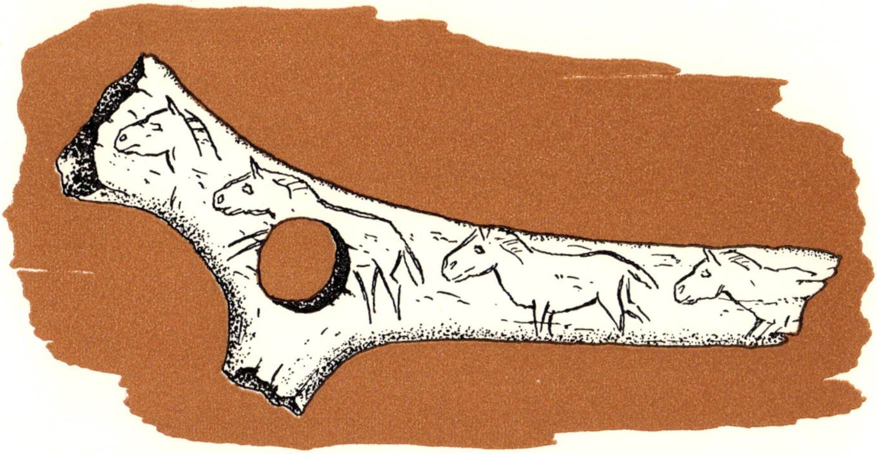

Cro-Magnons drilled holes in pieces of antler.

Certain objects which the Cro-Magnons carved out of reindeer antler are very puzzling. Nobody knows why they sometimes carved a design on a crooked piece of antler, and then drilled a neat round hole through it. Some experts think the cave men used such objects in pairs, as wrenches, to straighten the shaft of a spear. Others think they used these shaft-straighteners, as they are often called, for softening leather thongs. Still other experts think the curious objects were strung on leather thongs and worn around the neck, as ornaments or badges of honor.

Perhaps those crooked pieces of antler, and other carved objects the Cro-Magnons made—such as figures of women and of animals—are clues to their religion. Experts think the figures of women tell us that the Cro-Magnons worshipped a goddess as the source of all life. The animal carvings, some experts say, prove that the cave men were trying to please some god who could help them to be good hunters. But nobody can be really sure what the carved figures mean, or why the Cro-Magnons made them.

Most experts are also quite sure that the Cro-Magnons believed in a sort of magic—a magic that was a part of their religion. The experts think this is why those prehistoric men painted and carved pictures of animals on the walls and ceilings of certain caves. But even though we don't yet know for certain the reason for those cave decorations, it is easy to understand why people go to Europe from all over the world to see them. They are great works of art. When the first ones were discovered, almost nobody could believe that they had been created by prehistoric men who lived thousands of years ago.

11

A Little Girl and the Altamira Cave

A five-year-old girl discovered the amazing paintings in the cave called Altamira, in northern Spain. The story of her discovery, and what happened afterward, is one of the most famous of all stories about tracking down clues to prehistoric cave men.

The little girl's name was Maria, and she was the daughter of a Spanish nobleman, Don Marcelino de Sautuola. The family lived in a gray stone castle in the tiny village of Santillana del Mar, among the grass-grown limestone hills that rise up out of the sea along the coast of the Bay of Biscay.

The story really begins in 1878, when Don Marcelino visited Paris. There he saw a collection of prehistoric stone tools and carved antlers which had recently been found in France. He was so fascinated that he decided to make a hobby of searching for clues to prehistoric men. He also decided that a cave near his own castle would be a good place to begin his search. Northern Spain was not very far from the area in France where many prehistoric cave homes had already been discovered.

Before Don Marcelino left Paris, he asked all sorts of questions about the proper way to explore a cave. He learned that he would probably have to dig through many feet of piled-up earth before he reached a layer of genuine cave-man litter. He was told to do his digging very carefully, and to sift every shovelful of soil, so that he would be sure not to miss the tiniest scrap of bone or the smallest bit of chipped stone. He returned home determined to follow all the instructions very carefully. He thought he would be very proud if he could find even a few stone tools or bits of bone to add to the world's knowledge of man's prehistoric ancestors.

The only entrance to the cave which Don Marcelino had decided to explore was a narrow crack in the earth in a rocky meadow. No one had even known about the crack, or the cave itself, until a few years before, when a hunter's dog fell into it and disappeared. At that time the crack was covered over so that children could not crawl into the cave and be injured or lost.

But now, when Don Marcelino reached home, he reopened the cave entrance and went down

Don Marcelino opened up the old cave.

a sloping passage. At the foot of it was a cave room of fairly good size, with a passage leading off it only a few feet high. The passage was too low to work in. So Don Marcelino started digging slowly away at the floor of the front part of the cave.

Months went by. He was beginning to think he had made a poor guess, and that no prehistoric men had ever lived in that particular cave. But one day in November, 1879, he found several stones neatly chipped into the shape of spearheads. That night he hurried home to show his discoveries to his family, his servants, his farm workers, and all his neighbors.

Little Maria saw the spearheads too. She couldn't understand why everybody thought the stones were so important. But she could see that her father was very excited, and she decided that the cave he went to every day must be a wonderful place. She told Don Marcelino she wanted to go with him.

At first Don Marcelino shook his head. He felt sure Maria would quickly grow tired of trying to amuse herself in a dark cave. But he was very fond of his little daughter, and it was hard for him to refuse her anything. So he finally agreed to take her into the cave if she promised not to disturb him at his work.

Of course Maria was just as disappointed in the cave as her father had known she would be. There was nothing to play with. It wasn't much fun to watch her father patiently scraping away in the dirt. After a time, with a candle in her hand, she crept into the low passage.

Soon the little corridor grew broader. But it was still so low that even a small child like Maria had to be careful not to bump her head on the rough ceiling as she moved along.

Suddenly, glancing up at that ceiling, Maria stopped still. The candle wavered in her hand. All around her, in the flickering yellow light, there seemed to be animals—huge beasts with curving horns and staring eyes.

"Father!" Maria screamed. "Bulls! Bulls!"

Her voice echoed and re-echoed against the stone.

Don Marcelino dropped his tools, snatched up a candle, and crawled into the little tunnel.

Maria was the first to see the Altamira cave paintings.

"I'm coming, Maria!" he called. "Don't be frightened."

But when he came within sight of Maria, he realized that she was quite safe and not frightened at all. She was pointing at the ceiling.

"Look, Father!" she cried. "Look!"

At first Don Marcelino couldn't understand what he was supposed to look at. Twisting about to peer up at the ceiling, so close above his own head, he could see at first only some black lines, and some splotches of red and yellow.

Then slowly, one after the other, the splotches and lines took on shape. Shifting this way and that, and finally lying flat on his back in order to see more clearly, Don Marcelino at last realized why Maria had shouted out that startled word "Bulls!" Some of the shapes did look like bulls—huge staring beasts with curving horns.

Don Marcelino blinked his eyes. The more he stared, the more real the animals seemed. He realized that he was looking at paintings so skillfully done that the red and black and yellow creatures appeared to be alive.

There was one great beast, with head down, hind legs crumpled under him, as if he had been struck a mortal blow the very moment before Don Marcelino first saw him.

Not far away was a red horse, with a black mane and a black tail. It looked as if it had just touched the ground after a leap through space.

Utterly amazed, Don Marcelino looked from one picture to another. Red animals had been painted over animals outlined in black. A red creature was half covered up by another animal. The second one was painted in black and yellow, with his legs drawn together as if he were about to spring.

How had these paintings come to be made on the ceiling of a passage so low that a grown man couldn't sit upright in it? Who had made them? And when? Don Marcelino was certain that no one but himself, and a few of the men who worked for him, had entered the cave since the opening had been discovered a few years before.

Even as he tried to puzzle out an explanation for the remarkable figures leaping and striding across the rock vault over his head, he began to see that the ones that looked like bulls were not really bulls at all. He studied them more carefully. Then suddenly he caught his breath in excitement. The bull-like animals were actually bison—horned and humped creatures that had disappeared from Europe thousands of years before.

Don Marcelino realized that the other figures on the ceiling were prehistoric animals too. There were wild horses, long-antlered reindeer, and huge wild boars, of a kind no living man had ever laid eyes on.

That was when Don Marcelino understood the importance of little Maria's discovery. That very evening he wrote a long letter about the cave paintings to scientists in the Spanish capital, Madrid.

He wrote that the paint on the ceiling of the cave looked fresh and new, and that a bit of the red even came off on his finger when he touched it. But, he said, no one skillful enough to paint these figures had ever entered the cave since its discovery a few years earlier. Besides, he wrote, the animals were prehistoric animals, and none of Don Marcelino's Spanish neighbors knew what such animals looked like. He ended his letter by saying that he felt sure the amazing pictures had been painted at the time when bison roamed the Spanish hills. They had been painted, he said, by the same men who hunted those bison—by the cave men of prehistoric times.

His letter brought a geologist hurrying to the little village of Santillana del Mar to inspect the cave

In the Altamira cave, prehistoric men

which Don Marcelino called Altamira. At first the geologist thought Don Marcelino must be trying to play some sort of joke on him. No prehistoric men, the expert said, could have been skillful enough to paint those pictures. But when he began to cross-examine Don Marcelino, the geologist began to change his mind. When he explored the cave himself, he found in it some more of the chipped stones Don Marcelino had found, along with some bones of prehistoric animals. So the expert decided

painted reindeer, boars, bison, and horses.

the amateur was right. Immediately the expert returned to Madrid to make a public announcement of Maria's discovery.

Maria and the little village in which she lived became famous. Every day people came to visit the Altamira cave. Even the King of Spain arrived to admire the paintings and to congratulate the little girl who had found them. Copies of the paintings were prepared for a great conference of scientists from all over the world, soon to be held in Lisbon, Portugal. All the scientists coming to the Lisbon meeting were invited to visit Altamira.

Then that meeting took place, and suddenly everything changed. The scientists in Lisbon said they didn't even have to look at the pictures to be sure that the story of Altamira was nothing but a hoax.

Once more Santillana del Mar became a quiet village. No one came to visit it after the experts had their say. The newspapers that had been applauding Don Marcelino and his little daughter began to make fun of them instead. Experts refused to answer Don Marcelino's letters, and refused to let him attend their meetings to tell them about Altamira.

Finally Don Marcelino sealed up the entrance into the cave.

Years passed. Maria de Sautuola grew up. After Don Marcelino died, she was one of the few people who even remembered the brief exciting weeks when Altamira had been the pride of Spain.

In the meantime, all over the Dordogne region and farther south at the foot of the Pyrenees,

French prehistorians were exploring caves and discovering more and more clues to the prehistoric cave men. As they dug down through the layers of debris that filled some of the caves, they found hundreds of prehistoric bones and stone tools. Sometimes their digging even exposed drawings scratched on the walls. But the drawings were so crude that nobody doubted that they could have been made by primitive cave men. In fact, these primitive drawings were exactly what most people expected of prehistoric cave dwellers. Nobody made any connection between the crude scratchings and the paintings which the King of Spain had once admired as works of art.

Then, in 1901, French experts began to clear away the debris choking two caves close to the village of Cro-Magnon, the village where young Louis Lartet had first gazed at those five skeletons of Cro-Magnon man. Each inch of litter and soil they dug out revealed another inch of cave wall. By the time they reached the original floors of the two caves, the experts knew they had dug away debris that had lain undisturbed in those caves for many thousands of years. The walls exposed by the digging had been covered up all that time. And the walls were decorated!

On the walls of the long twisting cavern called Les Combarelles were hundreds of figures of prehistoric animals scratched into the rock—animals that seemed to run and leap and lunge like living beasts.

On the walls of the Font-de-Gaume cave, there were only a few paintings in red and black. But everybody who saw them agreed that the red

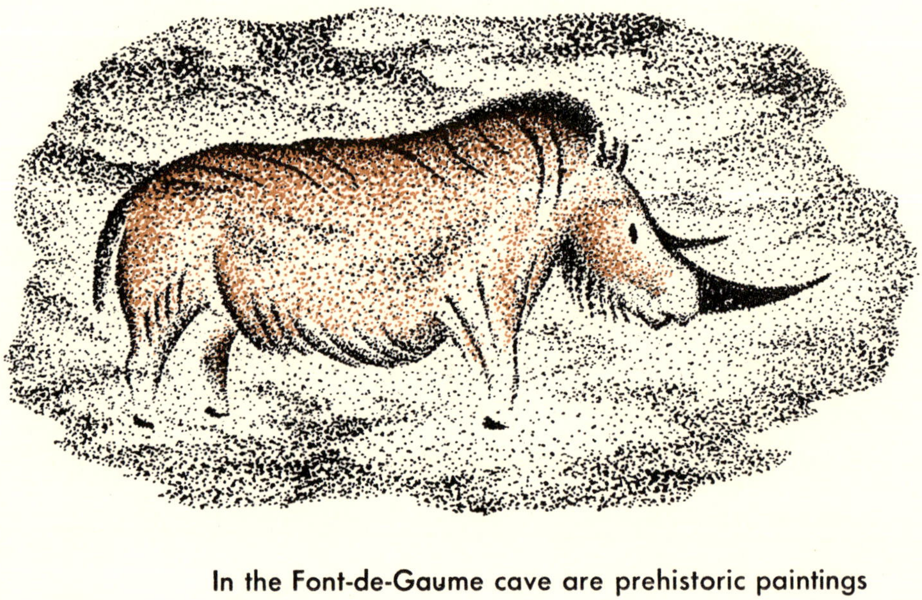

In the Font-de-Gaume cave are prehistoric paintings

woolly rhinoceros and the black-and-red reindeer looked alive too.

And all the experts who had been present while the caves were being cleared knew that the decorations had been put on the walls before the caves became filled up. Each man had to face the fact that those painted and carved walls proved two startling things.

They proved that the cave men had discovered the idea behind all art—the idea that lines and colors, on a flat surface, can give the impression

of a woolly rhinoceros (left) and a reindeer (right).

of an animal, or of some other real and solid object.

The paintings and drawings also proved that the cave men were really good artists. Some of the pictures were so simple that they looked rather like the work of a small child. But in almost every case the animals on the cave walls seemed alive. Only a really skillful artist can give the appearance of life to a group of lines and splotches of color.

One of the experts who gazed in astonishment at the freshly exposed works of art in those two French caves was a famous French paleontologist, Emil Cartailhac. And as he stood before them he remembered that he had flatly refused, almost a quarter of a century earlier, to visit a certain Spanish cave called Altamira. He had refused because he couldn't believe that what its discoverer said about it was true. Now Cartailhac realized that the Spanish amateur must have been right after all.

Cartailhac was an honorable man. He immediately made a public statement accusing himself for having doubted Don Marcelino's word without even going to look at the Altamira paintings himself. Then he packed his bags and set out for Spain.

A few days later Maria de Sautuola led the famous Frenchman up to the sealed doorway of Altamira. She turned the key in the rusty lock, pushed the door open, and took him inside. Quietly she lit candles and showed him the pictures that she had first seen when she was five years old.

Cartailhac could scarcely believe his eyes. The paintings of Altamira had been so marvelously preserved, in the cool cave, that they looked as if they had been made only an hour before. But Cartailhac, the expert, had no difficulty in recognizing their vast age. And he knew that they were by far the greatest example he had ever seen of the art of prehistoric man.

From that day on, the village of Santillana del Mar once more became a place of pilgrimage, crowded with eager visitors. Experts came from every quarter of the globe. Those who couldn't come urged that copies of the pictures be made, so that they too could see the paintings that had suddenly become so famous.

A handsome book was published, showing all the hundreds of drawings and paintings on the ceiling of Altamira. The floor of the cave was dug away so that the works of art could be seen more easily. Eventually the Spanish government took over responsibility for the cave, so that its precious paintings, one of the greatest art treasures of Spain, would always be protected. Every year, Altamira becomes more famous, and more people go there to study and admire the pictures little Maria de Sautuola first saw by the light of her candle.

Maria de Sautuola lived to hear all the experts admit that her father had been right. She lived to hear her own name spoken with respect and admiration throughout the world. She had discovered the clue that proved beyond a doubt that some of the greatest artists of all time were cave men who lived among bison and mammoths in the prehistoric period called the Ice Age.

12

Four Boys and the French Cave of Lascaux

Four boys discovered the wonderful prehistoric cave paintings of Lascaux, in the Dordogne region of France. Their story is as famous as the story of little Maria de Sautuola and the Altamira cave. It began one morning in September, 1940, during the Second World War, in the town of Montignac.

Montignac is on the Vézère River, less than twenty miles from Cro-Magnon. Limestone cliffs rise up all about it. Among the cliffs are several caves which, by 1940, prehistorians had already explored. Like almost everybody in the Dordogne region, the boys knew about those caves. They had

The Dordogne cliffs contain many caves.

visited the nearest ones with their teacher, and listened carefully when he told them how to recognize ancient bones and stone tools. They hoped to be lucky enough themselves to find some new clues to prehistoric cave men.

On the morning their story begins, however, the boys weren't thinking about prehistoric times. France had been conquered by Nazi Germany just three months before, and the usually cheerful town of Montignac had become a sad and dismal place.

When the boys wanted to escape it for a few hours, they wandered among the forested cliffs along the river. That's why they set off for the cliffs on the morning of September 12, with a little dog named Robot who belonged to one of them.

Suddenly, while the boys were climbing the hill called Lascaux, one of them noticed that Robot was no longer at their heels. Marcel, the dog's owner, whistled for him and called out his name. But the frisky dog didn't come bounding back to his place. Marcel's friend, Jacques, and the other boys called Robot too. Still he didn't appear, or answer them with his bark. Finally they decided he must have been hurt, and they turned back anxiously to look for him.

Every few seconds they stopped to whistle and shout, and to listen for some kind of response. At last they heard a frenzied barking. But it sounded so faint and so far away that they couldn't imagine where the dog was.

They called again. Again the bark replied. This time they realized that the sound came from under their feet, from somewhere underground. A moment later they found a small hole, not much larger than a rabbit burrow. They realized that Robot had somehow fallen down it and couldn't get out.

Swiftly the boys fell to work to enlarge the opening. They cut away the roots of shrubs with their

The dog disappeared down a hole.

knives, and pried out small stones, calling to Robot all the while to assure him that he would soon be free. But the little dog didn't appear even when the hole had been doubled in size. His barking didn't seem to come any closer.

"I'll have to go down and get him," Marcel said.

So the boys made the hole bigger still, until Marcel could squeeze himself into it.

"Don't worry, Robot," he called. "I'm coming."

Marcel pushed himself into the opening head first, and worked his way for several yards along a steeply slanting tunnel. When he thought he had reached the bottom, and tried to get to his feet, he lost his balance and found himself rolling helplessly downward. His journey came to an end with a thud that jarred his bones.

The other boys heard the rattle of stones and the crash of their friend's landing. "Are you all right?" Jacques shouted.

Marcel shook himself. He felt bruised, but otherwise he was uninjured. And Robot was leaping about him, licking happily at his hands and face.

"I'm all right, and so is Robot," he called upward. "But it will be a hard climb to get out again."

"What's down there?" the other boys wanted to know.

Marcel flicked the switch on the small flashlight that had been in his pocket. "A cave, I think," he said. And then he added excitedly, as he swung the beam around, "Yes—a cave! Come on down! But be careful or you'll fall the last part of the way."

One after another the other boys slithered down, clutching at roots and rocks in an effort to slow their swift progress. Finally they all stood together on the bottom, brushing themselves free of dirt while Robot barked his welcome.

"Look!" Marcel told them, pointing his flashlight. "Over there!"

Only a few feet away an enormous animal loomed out of the surrounding darkness, bright red against a paler background. Then they saw another animal, and still another.

After a startled moment, the new arrivals realized that the beasts were figures painted on the wall of

The boys saw an animal painted on the cave wall.

Prehistoric artists had painted the heads of a row of deer.

the cave in striking black and vivid yellow and red.

"They're like the pictures at the Font-de-Gaume cave!" one boy whispered in awe. "Only there are so many more of them!"

"Let's explore!"

With Marcel in the lead, they moved forward slowly into the center of a great hall-like cavern almost a hundred feet long. The pictures were all around them, on both long walls and even stretching up onto the arched ceiling.

There were prancing wild horses with tousled

manes. There were prehistoric bulls—four of them, three times larger than life, sprawled across the ceiling. There were bison and long-antlered deer. One row of deer heads looked exactly like a group of animals swimming across a stream—across the Vézère River itself, perhaps—with only their heads showing above the water.

It was late when the boys reluctantly scrambled back up the steep slope, taking Robot along with them. By then they knew that they had made a discovery so important that they would have to show it to other people, their teacher first of all.

But they had already promised each other that they would keep their exciting find a secret until they had a chance to explore it completely themselves. Cautiously they made their plans.

The next morning the boys left Montignac one by one, so that no one would become curious as to the group's purpose. Each boy carried a flashlight in his pocket. They met at the entrance to the cave and again Marcel led the way down the sloping tunnel to the cavern thirty feet below the surface.

That day they explored not only the big hall at the foot of the slope, but the two passages leading out of it at its far end. One passage ran straight forward. The other ran to the right, and divided into two smaller branches after a short distance.

The walls of both the passages were painted too. Altogether there were so many pictures, some painted on top of others, that the boys couldn't begin to count them all. There were pictures of single animals, some bigger than life, like the bulls on the ceiling of the hall, others scarcely larger than a man's hand. There were groups of animals shown together. There were animals that seemed to be standing still, and others that seemed to be leaping or falling or racing. Here and there were criss-crossed lines that looked as if they might be pictures of animal traps, and long straight lines that looked like spears.

The strangest picture of all was right at the bottom of a deep pit at the end of one of the passages. It was strange because there was a man in it, drawn in single lines like a stick figure drawn by a small child. He seemed to be hurt, for he was toppling backward. On one side of him was a wounded

One painting showed a bison, a bird, and a man.

bison with wildly staring eyes looking right at the boys as they crouched at the bottom of the pit. Perched on a stick on the other side of the man was a bird, crudely drawn, but looking very much alive, as if he were watching over the death of the beast and the man.

The next day and the day after that, the boys returned to their cave. At each visit they found something they hadn't noticed before, and grew more excited than ever about their own secret prehistoric art museum. But they knew they couldn't keep their secret forever. On the fifth day they told their teacher what they had found. At first the man thought they must be joking. Even when he was convinced that the boys had actually discovered a prehistoric painted cave, he couldn't believe that it was as remarkable as they insisted it was. But finally he agreed to go and look at it himself.

The teacher didn't speak at all when he first stared around the big painted hall. He just gasped in amazement. Then he said, "We must waste no time. We must send at once for Abbé Henri Breuil."

The boys knew that name. They knew that the

The teacher was amazed by what he saw.

French priest Abbé Breuil had been chosen to copy the famous paintings at Altamira. They had heard the story of how the young man had lain on his back for weeks, under the low Altamira ceiling, in order to set down every line and every shade of those paintings. The copies he had made of them, published in a book, had introduced the whole world to the glories of that Spanish cave.

Since that time Abbé Breuil had become

France's—perhaps the world's—greatest expert on prehistoric art. If Abbé Breuil decided that the paintings the boys had discovered were worth copying too, then the cave on Lascaux hill would also become famous throughout the world.

Abbé Breuil came at once. Not even the difficulties of transportation in wartime France could delay him. And after his first words the boys knew that the town of Montignac would never seem such a sad and dismal place again. They knew that from then on there would always be something exciting happening on the Lascaux hill.

For Abbé Breuil, and the other experts who soon joined him in Montignac, agreed that the cave was one of the greatest art treasures in France. They said it must be opened to the public so that all the world could marvel at the paintings. But the experts also agreed that the priceless paintings might be ruined if the cave were opened to the public without careful preparation.

Those men knew that the paintings had survived for thousands of years only because the air in the sealed-up cave had remained at the same degree of temperature and moisture all that time. If the cave were opened and outside air poured through it, those conditions would change and the pictures might soon begin to fade and peel off the walls.

So they sent for engineers to plan a special air-conditioning system to protect the prehistoric art. Other engineers made plans to enlarge the cave's entrance and provide it with massive airtight doors. Still other technical experts designed a lighting system that would show the paintings to their best advantage. Until 1945, when the war ended and France became free once again, very little actual work could be done. Even after the war it took a long time to build the special air-conditioning and ventilating machinery that was needed.

Finally, in 1948, everything was finished. The famous Lascaux cave was opened to public view, and people began to arrive at Montignac from all over the world. Two young guides were waiting to show them through the Hall of the Bulls, as the big cavern had been named.

The guides had been carefully trained for their job, and could tell the visitors many things about the Lascaux paintings. They could point out, first of all, how cleverly the artists had used the roughness of the rock. By drawing the outline of a bull around a bulge in the wall—instead of drawing it on a smooth flat place—a prehistoric artist had made a picture that seemed to have thickness as well as height and width. The bulging rock made the animal look real and alive.

The guides could also explain that the cave men probably made their paints by mixing certain substances with animal fat. Charcoal from a fire, the guides said, gave the artists their black paint, and powdered minerals gave them their red and yellow colors.

The guides could explain the painting methods the artists used too. The earliest cave men simply dipped their fingers in paint and used their finger tips for drawing lines on the stone. Later artists probably used brushes made out of feathers. Some of the prehistoric painters also used a method something like our modern spray painting. They spread fat on the wall, and then blew powdered color at it through a hollow reed. The color stuck to the greasy surface and formed a layer of paint that had a light misty look.

The guides told the visitors that no one knew

Some prehistoric artists blew colors onto the wall.

for certain why prehistoric men—who apparently never decorated the shallow shelters and cave fronts where they lived—had made hundreds of pictures deep inside Lascaux and certain other caves. But most experts agreed, the guides said, that the pictures were made because the cave men believed in magic. When a cave man went into a dark corner of a cave and drew a bison with a black line like a spear piercing its body, he was—the experts believed—trying to work a magic spell that would make him a successful hunter.

The guides could tell the visitors all those things because they had learned them from scholars like Abbé Breuil. But when the young men described the discovery of Lascaux, they were telling a story they knew better than any of the experts did.

The two guides who were on hand in 1948 to show the wonders of the Lascaux cave were Marcel Ravidat and Jacques Marsal. Eight years before, on an autumn day in 1940, they had been two of the four teen-aged boys who had followed little Robot down a hole and found, deep in the earth, the greatest collection of prehistoric cave paintings yet known to the world.

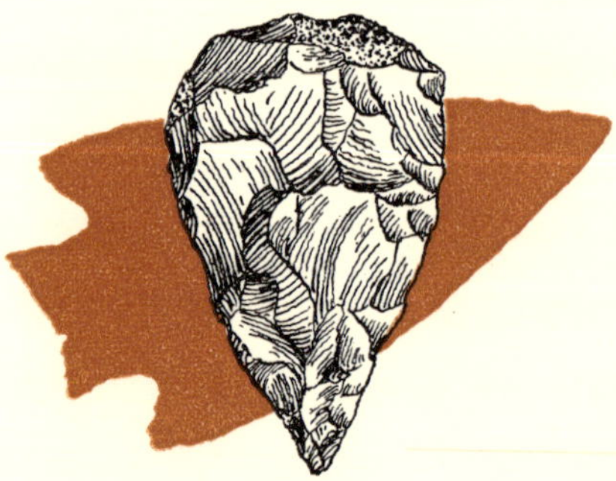

13

More Clues Are Needed

Man's search for clues to his prehistoric ancestors has been going on for only a little more than a hundred years. This search has already solved many mysteries of the past. And now our knowledge of that past is growing more rapidly than ever. Every year more people are searching for clues. And more people are trained to recognize the clues that are found, and to fit them into a pattern.

Recognizing clues and fitting them into a pattern are jobs for experts—the kind of experts that were unknown in the days of Jacques Boucher de

Perthes, before the science of prehistory existed. But if De Perthes were alive today he would not have to remain an amateur scientist. He would be able to study, under trained experts, and become an expert himself. And today he would have the help, in his work, of scientists in many other fields. Modern physicists and chemists, along with archeologists and geologists and paleontologists, give important aid to modern prehistorians.

But all the experts, no matter how well-trained they are, must have many clues to study. That single tooth of Peking man, for example, bought in the Chinese drug shop, could not tell even the most brilliant modern prehistorian very much about the man to whom it once belonged. Hundreds of clues—teeth and bones and tools—had to be found in the Choukoutien caves before it was possible to learn very much about Peking man and his life. And there are still many mysteries about Peking man which will be solved, if they are ever solved at all, only by the finding of still more clues.

That's why so many modern scientists of prehistory spend years of their lives, just as De Perthes did, poking about in holes in the ground, patiently digging in the earth.

Even today, however, not all the important clues are found by experts. The quite recent discovery of the Lascaux cave was made by amateurs, and very young amateurs they were. They knew enough not to disturb what they had found, and to show it first to someone who would know how to understand and protect their find. Even amateurs today know the importance of leaving all such discoveries untouched, until experts can study them at the

Amateurs can make important discoveries.

very spot where they have been found. This knowledge, so widespread today, is one reason why the next century will probably unfold far more of the story of prehistoric man.

An important new discovery, made perhaps by an amateur, may be reported in your newspaper tomorrow or next week or next month. It may prove to be just the clue prehistorians have been hoping to find. Or it may upset one of their theories entirely, and send them off on a new track.

If you make a habit of keeping your eyes open, and of looking for such things as odd-shaped stones, you might make such a discovery yourself.

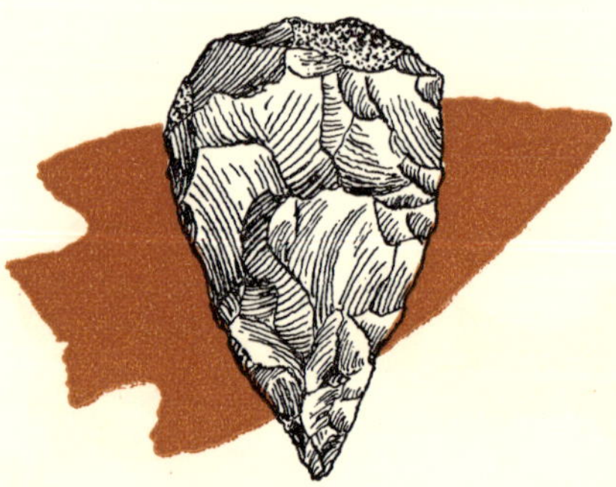

INDEX

A

B

C

K

L

M

N

P

Q

R

S

T

V

Z

The Authors

For *All About Prehistoric Cave Men,* Sam and Beryl Epstein traveled to England, France, and Spain, where they studied prehistoric art and visited many caves. In the course of their research, they estimate that they drove several thousand miles and walked at least a hundred—about a third of them underground.

The Epsteins are the authors of *All About the Desert* and more than fifty other books. They live in Southold, Long Island, N.Y.

The Artist

Will Huntington lives on the island of Martha's Vineyard, off the coast of Massachusetts. In addition to sailing and fishing, he likes to hunt for artifacts of the Indians who lived on the island before the coming of the white men. He has previously illustrated *The Tree of Culture,* a book about the history of man.

A-1 **All About Dinosaurs** *by* Roy Chapman Andrews
A-2 **All About Radio and Television** *by* Jack Gould
A-3 **All About the Sea** *by* Ferdinand C. Lane
A-4 **All About Volcanoes and Earthquakes** *by* Frederick H. Pough
A-5 **All About the Weather** *by* Ivan R. Tannehill
A-6 **All About the Insect World** *by* Ferdinand C. Lane
A-7 **All About the Stars** *by* Anne Terry White
A-8 **All About Whales** *by* Roy Chapman Andrews
A-9 **All About the Wonders of Chemistry** *by* Ira M. Freeman
A-10 **All About the Atom** *by* Ira M. Freeman
A-11 **All About Birds** *by* Robert S. Lemmon
A-12 **All About Our Changing Rocks** *by* Anne Terry White
A-13 **All About Rockets and Jets** *by* Fletcher Pratt
A-14 **All About the Flowering World** *by* Ferdinand C. Lane
A-15 **All About Moths and Butterflies** *by* Robert S. Lemmon
A-16 **All About Snakes** *by* Bessie M. Hecht
A-17 **All About Strange Beasts of the Past** *by* Roy Chapman Andrews
A-18 **All About Famous Inventors and Their Inventions** *by* Fletcher Pratt
A-19 **All About Strange Beasts of the Present** *by* Robert S. Lemmon
A-20 **All About the Arctic and Antarctic** *by* Armstrong Sperry
A-21 **All About the Desert** *by* Sam and Beryl Epstein
A-22 **All About Great Rivers of the World** *by* Anne Terry White
A-23 **All About Electricity** *by* Ira M. Freeman
A-24 **All About Famous Scientific Expeditions** *by* Raymond P. Holden
A-25 **All About Animals and Their Young** *by* Robert McClung
A-26 **All About Monkeys** *by* Robert S. Lemmon
A-27 **All About the Human Body** *by* Bernard Glemser
A-28 **All About Satellites and Space Ships** *by* David Dietz
A-29 **All About the Jungle** *by* Armstrong Sperry
A-30 **All About Prehistoric Cave Men** *by* Sam and Beryl Epstein
A-31 **All About the Ice Age** *by* Patricia Lauber
A-32 **All About Archaeology** *by* Anne Terry White